ZHANG XIANGQIAN'S UNIFIED FIELD THEORY (*POPULAR SCIENCE EDITION*)

EXTRATERRESTRIAL TECHNOLOGY

Hope Grace
PUBLISHING
Alexandria, Virginia, USA

By Lynn Lou Beran

Published by Hope Grace Publishing 2025
HopeGracePublishing.com
Alexandria, Virginia, USA

This book is an adaptation of the original scientific theories developed by **Zhang XiangQian**, presented in a popular science format by **Lynn Lou Beran**. While all efforts have been made to maintain the integrity of the original concepts, this edition is designed to make these theories more accessible to a general audience.

ISBN: 978-1-966423-06-5 (Paperback)
ISBN: 978-1-966423-12-6 (Hardback)
ISBN: 978-1-966423-07-2 (eBook)

Library of Congress Control Number: 2024925089

First Edition: 2025

Contents

Preface

In a world where science often feels out of reach to those without advanced degrees, it is essential to remember that some of the greatest scientific breakthroughs have come from unconventional thinkers. Zhang XiangQian's *Unified Field Theory* is one such breakthrough, offering a radical new way to understand the forces of nature and the fabric of the universe. It challenges traditional models and presents a unified explanation of gravity, electromagnetism, and nuclear forces — all through the lens of space's dynamic motion.

This *Popular Science Edition* was created to bring Zhang's groundbreaking work to a wider audience. By refining and simplifying the core concepts, this book aims to make the ideas within Zhang's *Unified Field Theory* accessible to readers of all backgrounds. Whether you're a curious mind with a love for science or someone seeking to understand the deeper mysteries of the universe, this edition is designed for you.

I have worked closely with Zhang XiangQian to bridge the gap between advanced theoretical physics and everyday understanding. My journey with this theory began with translating the first edition word-for-word, and through that process, I realized the immense potential Zhang's ideas have to reshape our understanding of reality.

This edition covers not just the foundational principles of *Unified Field Theory* but also explores its potential applications in energy, space travel, and even consciousness. We believe these concepts will ignite a sense of wonder in readers and inspire future generations to explore new frontiers in science and technology.

We invite you to dive into the world of space, motion, and forces and to imagine the vast possibilities that Zhang's theory presents. As you turn these pages, you'll embark on a journey through the universe — one where the limits of science are redefined, and the potential for human advancement is boundless.

Thank you for joining us on this exciting adventure.

Introduction: Unveiling the Secrets of the Universe

What if the key to understanding the universe lies not in the stars, but in the very space that surrounds us? What if space itself is more than an empty void — constantly in motion, and fundamental to the nature of mass, energy, and light? These are just some of the groundbreaking questions explored in Zhang XiangQian's *Unified Field Theory*.

Zhang XiangQian's theory offers a revolutionary new perspective — one that challenges long-standing principles in physics and suggests groundbreaking ways to manipulate **space, mass, and light-speed travel**. Central to this theory is the concept that **mass is not an intrinsic property** of matter but a result of the **interaction with space itself**. By understanding the **cylindrical spiral motion of space**, Zhang unlocks new possibilities in **mass reduction**, allowing objects to travel at the speed of light by **reducing their mass to zero**.

This **Popular Science Edition** makes Zhang's complex theory accessible to those without a deep background in physics. While the ideas may seem bold and speculative, they are rooted in experimental insights that suggest the real-world potential of these concepts. The theory goes beyond conventional thinking and offers a glimpse into a future where technology could be revolutionized by manipulating space itself.

In this book, we'll explore:

1. Faster-than-Light Communication

How manipulating space through cylindrical spiral motion could allow communication across vast distances with minimal energy loss, potentially overcoming the limits of traditional electromagnetic waves.

2. Light-Speed Travel through Mass Reduction

The secret to achieving light speed by reducing mass to zero, a key element of Zhang's theory, and how this principle could revolutionize space travel.

3. Instantaneous Travel

The potential of space manipulation to enable instantaneous travel across vast distances, rethinking the boundaries of transportation and movement.

4. Artificial Fields for Matter and Energy Control

The possibility of creating artificial fields to manipulate matter and energy, allowing for advancements in virtual structures and material manipulation.

5. Body Replication and Consciousness Transfer

How advances in our understanding of space and energy could lead to technologies that allow the replication of bodies or even the transfer of consciousness, reshaping our conception of life and existence.

Through the **unification of the four fundamental forces** — gravity, electric force, magnetic force, and nuclear force — Zhang's theory offers revolutionary implications for energy production, space travel, and our understanding of time. This edition not only simplifies Zhang's groundbreaking work but also provides real-world analogies and speculative applications that stretch the imagination. Whether you're a curious reader, an aspiring scientist, or someone fascinated by the future of technology, this book is your gateway to an exciting new frontier in physics.

Part 1: The Foundations of Space and Motion

Chapter 1: What Is Space, Really?

When most people think of space, they picture an empty void—a vast nothingness that stretches between stars, planets, and galaxies. But according to Zhang XiangQian's *Unified Field Theory*, space is much more than just emptiness. In fact, space is a dynamic, ever-moving entity that plays a crucial role in the nature of mass, energy, and light.

1. The Traditional View of Space

For centuries, scientists considered space to be a static background — a kind of stage on which the events of the universe played out. It was something that contained objects but didn't actively interact with them. Einstein's theory of general relativity changed this perspective by introducing the idea that space is not only affected by gravity but can also be bent and curved by massive objects (Einstein, 1915).

However, Zhang's theory goes a step further. Rather than seeing space as merely reactive — bending in response to mass — Zhang proposes that space is in **constant vectorized motion**. This motion is fundamental to the nature of objects themselves, extending beyond the concept of scalar speed to include directionality, offering new ways to manipulate space for controlling mass and energy.

2. The Cylindrical Spiral Motion of Space

Zhang XiangQian introduces a radical idea: space itself moves in a **cylindrical spiral pattern** around every object in the universe. This motion is not just theoretical; it is the core principle that explains how objects exist and interact with the forces around them. Imagine space as a fluid constantly flowing outward from every object at the speed of light in a vectorized direction. This means that both the speed and the direction of space's movement play a critical role in how objects are affected (Zhang, 2025).

3. Space as a Dynamic Entity

In Zhang's theory, space is not passive; it is actively involved in the properties of matter. The movement of space, particularly its **cylindrical spiral vectorized motion**, directly affects the mass of an object. This differs from traditional physics, where mass is treated as an intrinsic property of an object, independent of space. By rethinking space as an active player in the universe, Zhang challenges the long-held notion that space is a mere container for physical events.

Instead, space is a key factor in determining the behavior of mass and energy. This dynamic view of space sets the stage for understanding more advanced concepts in later chapters, such as **mass reduction** and **light-speed travel**.

4. How Does This Change Our Understanding of the Universe?

If space is in constant motion as Zhang suggests, then every object in the universe is embedded within this motion. This means that the way objects interact with each other isn't just due to forces like gravity or electromagnetism, but also due to the vectorized motion of space itself. This concept allows for greater precision in understanding and manipulating space.

This reimagining of space has profound implications. It suggests that space is not only fundamental to the way objects behave, but is also the key to manipulating mass and energy. Understanding this constant motion, particularly its **vectorized nature**, could pave the way for future technologies that exploit space's behavior to achieve feats like **light-speed travel** or even **teleportation**.

5. Conclusion

Zhang XiangQian's theory of space as a moving, dynamic entity challenges the conventional understanding of the universe. By introducing the idea of **vectorized cylindrical spiral motion**, Zhang offers a new framework for understanding how mass, energy, and space interact. This chapter lays the groundwork for the more advanced topics to come, including the manipulation of mass and energy, the unification of forces, and the potential for new revolutionary technologies.

Chapter 2: Why Space Surrounds Every Object

In Zhang XiangQian's *Unified Field Theory*, there are two foundational entities in the universe: the **object** and the **space surrounding it**. These two entities form the basis for understanding mass, energy, and the forces of nature. According to Zhang, the space around an object is not passive; it is constantly in motion, shaping the properties of the object itself.

1. The Role of Space in Mass

Mass, in Zhang's theory, is not an intrinsic property of an object but rather a result of the space surrounding it. This space is constantly moving in a **cylindrical spiral pattern**, radiating outward at the speed of light. The interaction between an object and this dynamic space determines how we perceive its mass. In other words, mass is not something an object inherently possesses; it is something that space generates for the object.

In Zhang's *Unified Field Theory*, **mass** is defined by how space moves around an object. More specifically, mass is determined by how many small movements of space, called **spatial displacement vectors**, are packed into the space surrounding the object. The equation that describes this is:

$$m = k \frac{\oiint dn}{\oiint d\Omega} = k \frac{n}{4\pi}$$

In simpler terms, this means that the mass of an object depends on how these space movements are distributed in all directions around the object, within a full spherical angle of 4π. The movement of space at any point around the object is linked to the speed of light, which is represented by $\vec{r} = \vec{c}t$.

By recognizing the **object** and the **surrounding space** as the two most foundational entities, Zhang's theory opens the door to manipulating mass by controlling the space around an object. This concept of **mass reduction** forms the basis of potential breakthroughs like **light-speed travel** and **instantaneous transportation** (Zhang, 2025).

2. Space Superposition: How Space Interacts with Itself

One of the key insights from Zhang's *Unified Field Theory* is that space around different objects can **superpose** or overlap. When this happens, the space surrounding each object interacts, influencing the distribution of **spatial displacement vectors** that determine mass. Mass is not an isolated property of an object; it emerges from how space behaves around it.

From Zhang's mass definition:

$$m = k \; \frac{\oiint dn}{\oiint d\Omega} = k \; \frac{n}{4\pi}$$

we understand that mass is a measure of the distribution of spatial displacement vectors around an object within a spherical angle of 4π. When two or more objects are close enough for their surrounding spaces to overlap, the **spatial displacement vectors** from each object's space can either **amplify** or **cancel out** depending on their alignment. This interaction between spaces is what Zhang refers to as **space superposition**.

Imagine two objects, each radiating a cylindrical spiral of space. If the spiral patterns of these spaces are aligned in a way that reinforces each other, the spatial displacement vectors around the objects will combine, increasing the mass effect. Conversely, if the spirals are aligned in such a way that they oppose each other, they will cancel out, reducing the mass.

In this scenario, **space superposition** offers a mechanism for **mass reduction**. When the space surrounding two objects superposes in a way that reduces the effect of the spatial displacement vectors, the total mass of the objects can be minimized. This principle allows for the possibility of reducing an object's mass to near zero without altering its physical structure, thereby enabling **instantaneous light-speed travel** or **massless transportation**.

3. Mass as a Result of Space Interaction

In traditional physics, mass is viewed as an inherent property of matter. Zhang XiangQian's theory, however, challenges this view. Mass is not something that exists independently within an object but is created by the space surrounding it. The motion of space — particularly its cylindrical spiral motion — creates what we perceive as mass. By manipulating this motion, it is possible to change the mass of an object or even reduce it to zero.

This understanding of mass as a dynamic interaction between an object and space provides new opportunities for controlling mass. Instead of viewing mass as a fixed property, Zhang's theory suggests that by adjusting the space surrounding an object, we can alter its mass. This opens the door to technologies that could reduce the mass of objects, making them easier to move or even allowing them to achieve light-speed travel.

4. Why Space Surrounds Everything

Zhang's theory emphasizes that every object in the universe is surrounded by space that is in constant motion. This space is essential because it defines the object's properties. Without space, objects would not have mass, energy, or even the ability to interact with one another. In this sense, space is as much a part of the object as the matter that makes it up.

Furthermore, the dynamic nature of space — its cylindrical spiral motion — suggests that objects are not isolated entities but are deeply connected to the universe through the space that surrounds them. This connection between space and object is key to understanding not only mass and energy but also the potential for **mass manipulation** and the extraordinary capabilities that could result from it.

5. The Implications of Space Superposition

The idea that space can superpose and interact with itself opens up a new world of possibilities. If we can learn to control the space surrounding objects, we could achieve remarkable feats, such as reducing the mass of heavy objects, making them easier to move, or even eliminating mass entirely for instant light-speed travel.

Space superposition also suggests that space is not just a passive container for objects. It actively shapes the properties of objects, determining their mass, energy, and motion. By controlling how space superposes and interacts, we can unlock new capabilities in transportation, energy systems, and even material science.

6. Conclusion

Zhang XiangQian's theory shows us that space is not just the empty void between objects. It is a dynamic, moving entity that plays a crucial role in defining the properties of everything it surrounds. By understanding the cylindrical spiral motion of space and how space superposition can lead to mass reduction, we gain new insights into the nature of mass and the universe itself.

One of the most revolutionary implications of mass reduction is that when mass is reduced to zero and an object reaches light speed, the distance between objects effectively shrinks to zero. This means that, at light speed, the separation between two points becomes irrelevant, and an object can travel instantly from one location to another. This concept bridges the understanding of mass manipulation with instantaneous travel, setting the stage for exploring the vast possibilities of light-speed transportation.

This chapter sets the foundation for how this understanding can be applied to achieve light-speed travel and revolutionize our approach to mass and energy.

Chapter 3: The Speed of Light and Its Role in the Universe

For centuries, the speed of light has been viewed as one of the most fundamental constants in the universe. At approximately 300,000 kilometers per second (186,000 miles per second), it is the fastest speed at which anything can travel. From the perspective of traditional physics, nothing with mass can ever reach the speed of light because the energy required would become infinite. However, Zhang XiangQian's *Unified Field Theory* introduces a groundbreaking idea: the speed of light is not just a limit but a **vectorized motion** that can be controlled, and by manipulating space, it's possible to achieve **instantaneous light-speed travel**.

1. The Traditional Understanding of Light-Speed Limits

In Einstein's theory of relativity, the speed of light is the ultimate speed limit. As objects with mass approach the speed of light, their mass increases, requiring more and more energy to accelerate. According to this theory, achieving light-speed for any object with mass would require infinite energy — something that is practically impossible. For decades, this concept has defined our understanding of the universe, limiting our hopes of faster-than-light travel or space exploration to distant stars.

However, Zhang XiangQian's theory challenges this assumption by proposing that mass is not fixed and that space can be manipulated in such a way that **light-speed travel** can be achieved without requiring infinite energy.

2. Vectorized Light-Speed Motion

One of Zhang's key contributions is the idea that the speed of light is not just a scalar quantity — a simple value that defines how fast something moves. Instead, it's a **vectorized motion**, meaning that both the speed and the direction of light-speed motion can be controlled. This concept is essential because it opens up the possibility of manipulating the direction of motion, not just the speed itself.

In Zhang's theory, space itself is in constant motion, moving outward from every object in a **cylindrical spiral pattern**. By controlling the vectorized motion of space, it becomes possible to influence how objects move within space. This idea forms the foundation for the potential of **instantaneous light-speed travel**, where the manipulation of space around an object allows it to reach the speed of light without the need for gradual acceleration or infinite energy (Zhang, 2025).

3. The Role of Space in Light-Speed Travel

Zhang's theory hinges on the idea that space is not just a passive backdrop but an active, dynamic entity that can be manipulated. The motion of space, particularly its **vectorized light-speed motion**, determines how objects with mass behave. By understanding and controlling this motion, it becomes possible to bypass the traditional limits imposed by physics.

In Zhang's view, objects are not confined by the traditional light-speed barrier because their mass is not fixed. Mass is the result of the space surrounding an object, and if the space can be manipulated, so too can the object's mass. By reducing the mass of an object to zero through **space superposition**, it becomes possible to eliminate the need for infinite energy and achieve **instantaneous light-speed travel**.

4. Instantaneous Light-Speed Travel: A New Frontier

Perhaps the most revolutionary idea in Zhang's theory is the concept of **instantaneous light-speed travel**. In contrast to traditional physics, which views the speed of light as an insurmountable barrier, Zhang proposes that if mass can be reduced to zero, an object can immediately reach the speed of light. There is no gradual acceleration required; instead, the moment mass is eliminated, the object moves at light speed.

This idea is based on the behavior of photons—particles of light that have no mass and always travel at the speed of light. By reducing an object's mass to zero through the manipulation of space, Zhang suggests that the object can behave like a photon, traveling at light speed without resistance. This opens the door to previously unimaginable possibilities, such as traveling to distant stars in an instant or even bypassing the traditional concept of distance altogether.

5. The Implications of Light-Speed Travel

If light-speed travel becomes possible through the manipulation of space, it would revolutionize our understanding of the universe and the way we navigate it.

The idea that distance can collapse to zero at the speed of light suggests that interstellar travel, once considered an insurmountable challenge, could become a reality.

In addition to the potential for space exploration, the ability to manipulate light-speed motion could have profound implications for transportation, communication, and even the structure of time itself. If distance becomes irrelevant at the speed of light, time could also become fluid, opening up new possibilities for human exploration and interaction with the cosmos.

6. Conclusion

The speed of light has long been considered the ultimate limit in the universe, a boundary that no object with mass could ever cross. But Zhang XiangQian's *Unified Field Theory* presents a new way of thinking about light-speed. By introducing the concept of **vectorized light-speed motion** and the possibility of **instantaneous light-speed travel** through the manipulation of space, Zhang opens the door to a future where the speed of light is no longer a barrier but a gateway to new possibilities. This chapter lays the foundation for understanding how mass reduction, space manipulation, and light-speed motion can work together to achieve what was once thought impossible: traveling at the speed of light.

Part 2: Breaking Down Mass and Energy

Chapter 4: The Nature of Mass – A New Perspective

For centuries, mass has been considered a fundamental property of matter, something intrinsic to objects that defines their resistance to motion and their interaction with forces such as gravity. In classical physics, mass is treated as something fixed, a quantity that belongs to an object and remains unchanged unless acted upon by external forces. However, Zhang XiangQian's *Unified Field Theory* offers a revolutionary new perspective: **mass is not an inherent property of objects**, but a result of the **motion of space** surrounding them.

1. Mass as a Result of Space Interaction

According to Zhang's theory, mass is generated by the interaction between an object and the space that surrounds it. This space is not static but moves in a **cylindrical spiral pattern**, radiating outward from the object at the speed of light. The interaction of an object with the space around it determines how much resistance the object has to motion, which is what we perceive as mass (Zhang, 2025).

In mathematical terms, **mass** can be defined by how space moves around an object. More specifically, it is determined by the distribution of **spatial displacement vectors** within the space surrounding the object, as described by the equation:

$$m = k \, \frac{\oiint dn}{\oiint d\Omega} = k \, \frac{n}{4\pi}$$

This equation means that mass is a measure of how many small movements of space (spatial displacement vectors) are packed into the space around an object, within a solid angle of 4π. These vectors represent the movements of space that contribute to the resistance, or mass, of an object (Zhang, 2025).

2. Mass as a Dynamic Property

In contrast to classical physics, where mass is treated as an intrinsic and unchangeable property of matter, Zhang's theory introduces mass as a **dynamic property** that depends on the behavior of space. By altering the way space moves around an object—through the manipulation of **space superposition** or other techniques — an object's mass can be reduced or even eliminated entirely.

This dynamic nature of mass opens up new possibilities for **mass manipulation**. If mass is a product of the surrounding space, then by controlling space, we can change the mass of an object without altering its physical structure. This forms the foundation for breakthroughs such as **mass reduction** and **light-speed travel**.

3. The Role of Space Superposition in Mass Reduction

Zhang's theory also emphasizes the role of **space superposition** in **mass reduction**. When the space surrounding different objects overlaps and interacts, the spatial displacement vectors can either combine to increase mass or cancel out to reduce it. This principle allows for the possibility of reducing an object's mass to near zero, which would enable the object to travel at the speed of light without requiring infinite energy.

In practical terms, **mass reduction** through space superposition could lead to revolutionary technologies, such as **massless transportation**, where objects could be moved with minimal energy, and **instantaneous travel**, where distances between two points effectively vanish as the object reaches light speed.

4. Mass and Energy: A New Relationship

In classical physics, mass and energy are closely related through Einstein's famous equation, E = mc^2, which shows that mass and energy are interchangeable. Zhang's theory builds upon this idea by suggesting that both mass and energy are not inherent properties of objects, but results of how space interacts with objects (Einstein, 1915).

In this sense, **mass** and **energy** are both products of the **spatial displacement vectors** that govern the motion of space. By controlling the movement of space, it is possible to influence not only mass but also energy, opening up new avenues for controlling and harnessing energy in fields such as transportation, power generation, and even material science.

5. Implications of a Dynamic Mass Model

The idea that mass is not a fixed property but rather a dynamic one has profound implications. It challenges long-standing principles in physics and suggests that with the right technology, we could control mass and energy in ways that were previously unimaginable.

Zhang's theory of **mass as a result of space interaction** not only redefines our understanding of mass but also provides a framework for technologies that can manipulate mass, energy, and motion. This could lead to advancements in **light-speed travel, energy-efficient systems**, and even **massless technology** that could transform industries.

6. Conclusion

Zhang XiangQian's *Unified Field Theory* redefines the concept of mass, showing that it is not an intrinsic property of objects, but a result of the interaction between objects and the space surrounding them. This new perspective on mass, combined with the idea of **space superposition** and **mass reduction**, opens up revolutionary possibilities for **light-speed travel** and **massless transportation**. As we explore the implications of this dynamic model of mass, we begin to unlock the potential for controlling mass and energy in ways that could reshape our understanding of the universe.

Chapter 5: Manipulating Mass – Is Light-Speed Travel Possible?

One of the most groundbreaking ideas in Zhang XiangQian's *Unified Field Theory* is the possibility of **mass reduction**, a concept that suggests we can reduce an object's mass to zero, allowing it to reach the speed of light without the need for infinite energy or gradual acceleration. This concept challenges conventional physics, which predicts that as objects with mass approach the speed of light, their mass increases to infinity. Zhang's theory offers an alternative: by manipulating the space surrounding an object, we can reduce its mass and make light-speed travel possible.

1. The Role of Mass Reduction in Light-Speed Travel

Zhang's theory redefines **mass** as the result of how space interacts with an object. As discussed in earlier chapters, mass is not an intrinsic property of matter but a consequence of the **spatial displacement vectors** that define how space moves around an object. The mass of an object is determined by the distribution of these vectors within the space surrounding it, described by the equation:

$$m = k \frac{\oiint dn}{\oiint d\Omega} = k \frac{n}{4\pi}$$

According to this definition, if we can manipulate the space around an object—specifically the spatial displacement vectors—it is possible to reduce the object's mass. When the spatial displacement vectors are canceled or minimized through **space superposition**, the object's mass approaches zero (Zhang, 2025).

As mass approaches zero, the object encounters less resistance to motion, allowing it to reach the speed of light almost instantaneously. There is no need for gradual acceleration; the object can transition directly to light-speed travel once its mass is eliminated. This breakthrough forms the foundation of **instantaneous light-speed travel** in Zhang's theory.

2. Space Superposition and the Elimination of Mass

At the core of **mass reduction** is the concept of **space superposition**. When the space around different objects overlaps, the spatial displacement vectors within that space interact. If the vectors are aligned in a way that cancels each other out, the object's mass is reduced.

By controlling how space interacts with itself through superposition, we can effectively manipulate the mass of objects. In a scenario where mass is reduced to zero, the object would behave like a photon—a particle of light with no mass—allowing it to move at the speed of light without requiring the enormous amounts of energy that traditional physics would predict.

This mechanism of **space superposition** is what enables **light-speed travel** without violating the principles of energy conservation. Rather than using immense amounts of energy to accelerate an object with mass, the solution is to reduce the object's mass through spatial manipulation, making it possible to travel faster than anything previously thought possible.

3. Instantaneous Light-Speed Travel

Zhang's theory introduces the possibility of **instantaneous light-speed travel** by reducing an object's mass to zero. In contrast to conventional physics, which suggests that objects with mass cannot reach light speed, Zhang's model shows that by eliminating mass, an object can bypass the limitations imposed by mass and energy.

Once the mass of an object reaches zero, the object is no longer subject to inertia or resistance. This allows it to achieve light speed instantly, without the need for gradual acceleration. The moment mass is reduced to zero, the object transitions to light-speed motion.

4. The Relationship Between Mass Reduction and Distance

A key implication of Zhang's theory is that when mass is reduced to zero, **distance shrinks to zero** as well. At light speed, the space between objects effectively collapses, meaning that an object can travel between two points instantaneously, no matter how far apart they are.

This concept challenges our traditional understanding of space and distance. From the perspective of an object moving at the speed of light, the distance between any two points in the universe becomes irrelevant. This suggests that **instantaneous travel** is not only possible but is a natural consequence of reducing mass to zero and reaching light speed.

5. Practical Challenges of Mass Manipulation

While the theory is sound, there are practical challenges to achieving mass reduction and light-speed travel. First, the technology required to manipulate space at such a precise level remains under development. Controlling the **spatial displacement vectors** and ensuring that they cancel out mass effectively is a complex process that requires a high level of precision.

Additionally, there is the question of **restoring mass** once light-speed travel is complete. After an object has traveled at the speed of light, its mass would need to be restored to ensure safe and stable motion at slower speeds. Managing this transition between zero mass and regular mass presents significant engineering challenges.

6. Implications for the Future of Space Travel

Despite these challenges, the potential applications of Zhang's theory are revolutionary. The ability to reduce mass to zero and achieve instantaneous light-speed travel could transform space exploration, making it possible to reach distant planets and star systems in a fraction of the time currently required.

In addition to interstellar travel, **mass manipulation** could also have profound effects on transportation within our own planet. **Instantaneous transportation** — the ability to move objects or people from one place to another without delay—could become a reality, eliminating the need for traditional forms of transportation like cars, planes, and trains.

7. Conclusion

Zhang XiangQian's *Unified Field Theory* offers a revolutionary approach to understanding mass, space, and motion. By redefining mass as the result of **spatial displacement vectors** and introducing the concept of **space superposition**, Zhang opens the door to the possibility of **mass reduction** and **instantaneous light-speed travel**. While significant challenges remain, the potential applications of this theory — both in space exploration and on Earth — are vast and groundbreaking.

While the technology to control space and mass remains in development, Zhang's theory offers a visionary glimpse into the future of human travel and exploration.

Chapter 6: Energy and Mass: Are They Truly Equivalent?

Energy has long been understood as a fundamental concept in physics, intertwined with mass and motion. In classical physics, energy is tied to the movement of objects and the forces acting on them. However, in Zhang XiangQian's *Unified Field Theory*, the concept of energy is redefined through the lens of **mass reduction**. By exploring how energy relates to the interaction of space with objects, Zhang provides a new perspective on how energy can be manipulated and controlled.

1. Energy and Its Relationship to Mass

In traditional physics, energy is closely linked to mass through Einstein's famous equation, $E = mc^2$, which shows that energy and mass are interchangeable. This means that as an object's mass increases, so does its energy. However, Zhang's theory challenges the idea that mass is a fixed quantity, proposing instead that **mass is the result of space interacting with an object**. Consequently, energy is also a dynamic property that can be influenced by manipulating the space around an object.

In Zhang's theory, **rest energy** arises from **rest momentum**, which is a special case of the **motion**

momentum equation when the object's velocity $\vec{v}$ is zero. Zhang's **signature equation for motion momentum** is:

$$\overrightarrow{p_{motion}} = m\,(\vec{c} - \vec{v})$$

where m is the mass of the object, $\vec{c}$ is the speed of light, and $\vec{v}$ is the velocity of the object. This equation reveals that an object's momentum is directly influenced by both its mass and the difference between the speed of light and its velocity.

When $\vec{v}$=0, the object is at rest, and the equation simplifies to:

$$\overrightarrow{p_{rest}} = m'\,\vec{c}$$

where m' is the object's rest mass. This equation demonstrates that **rest momentum** is the root cause of **rest energy** and shows how momentum remains fundamental even when the object is stationary (Zhang, 2025).

When mass is reduced through **space superposition,** the energy associated with that mass changes as well. By reducing mass, we are also reducing the energy needed to initiate motion. This principle is critical in Zhang's theory, where **mass reduction** not only enables light-speed travel but also opens up new ways to control energy.

2. Energy Conservation and Massless Travel

One of the most intriguing aspects of Zhang's theory is how **energy conservation** is maintained during **mass reduction**. In classical physics, an object's energy increases as it accelerates, requiring more and more energy to reach higher speeds. But Zhang's theory bypasses this need by reducing mass to zero, which allows an object to reach light speed without the infinite energy required in conventional physics.

As an object's mass approaches zero, the energy needed to initiate movement is minimized. Once mass is reduced to zero and the object reaches light speed, **no additional energy is required to maintain that motion**. This leads to the possibility of **instantaneous travel**, where an object's mass is reduced to zero in the first stage of travel, allowing it to reach light speed instantly and making the distance between two points shrink to zero. In the second stage, the object's mass is restored at the destination, ensuring stable operation at sub-light speeds.

This concept has profound implications for space travel and transportation. With mass reduced to zero, objects can move at light speed without requiring any additional energy for motion, offering a potential solution to the challenges of long-distance space exploration.

3. Energy as a Dynamic Property of Space

Zhang's theory shifts the focus from energy being solely tied to mass to energy being a **dynamic property of space itself**. Just as mass is defined by the **spatial displacement vectors** surrounding an object, energy is also influenced by the way space moves and interacts with matter. By manipulating space, it is possible to control not only mass but also energy, leading to more efficient forms of propulsion, power generation, and transportation.

In this context, **space superposition** plays a vital role. When the space surrounding objects overlaps and interacts, it influences both the mass and energy of the objects involved. By controlling this interaction, it is possible to reduce the energy required to start motion or even harness new forms of energy from the way space behaves.

4. Potential Technological Applications of Energy Control

The ability to manipulate energy through **mass reduction** and **space superposition** has far-reaching technological implications. For instance, spacecraft could be designed to **reduce their mass to zero, allowing them to instantly reach light speed**. As mass is reduced to zero, the distance between two points shrinks to zero, enabling instantaneous travel. Upon arrival at the destination, the spacecraft's mass would need to be **restored**, ensuring safe operation at sub-light speeds.

Additionally, energy systems could be developed that harness the movement of space itself to generate power, offering a more efficient and sustainable way to produce energy. This understanding that **energy is a dynamic property of space** could lead to innovations in energy storage and transfer, making it possible to store energy in new ways or even transfer energy across vast distances with minimal loss.

5. Energy and Light-Speed Travel

One of the most revolutionary aspects of Zhang's theory is its potential to overcome the traditional limits of energy consumption in space travel. In conventional physics, the amount of energy required to move an object increases dramatically as it approaches the speed of light. This presents a major obstacle to space exploration, as the energy needed to propel a spacecraft to light speed would be enormous.

However, Zhang's theory offers a solution by focusing on **mass reduction** as the key to light-speed travel. By reducing an object's mass to zero, the energy required to **initiate motion** drops significantly, and **no additional energy** is needed to maintain light-speed travel. In this state, the object can reach light speed and continue traveling without the need for continuous energy input, allowing for faster and more efficient travel across the universe.

6. Conclusion

Zhang XiangQian's *Unified Field Theory* redefines energy as a dynamic property of space that can be manipulated through **mass reduction, rest momentum**, and **space superposition**. The **signature equation for motion momentum** and the simplified form for **rest momentum** show how energy is fundamentally linked to momentum and mass. This new perspective opens up revolutionary possibilities for controlling energy, reducing mass, and enabling light-speed travel. Once an object reaches light speed, **no further energy is required**, reshaping our approach to space exploration, transportation, and energy generation. Zhang's theory offers a glimpse into a future where energy is harnessed in ways that were once thought impossible.

Part 3: Advanced Technologies for a New Era

Chapter 7: Faster-than-Light Communication

In the realm of modern technology, communication relies heavily on electromagnetic waves, including radio and light waves, to transmit information. These waves travel at the speed of light, making them the fastest known method for long-distance communication. However, even at light speed, transmitting messages across the vast distances of space takes significant time. For instance, it takes over four years for light to travel from Earth to the nearest star. In Zhang XiangQian's *Unified Field Theory*, the concept of **faster-than-light communication** becomes feasible through the manipulation of space itself.

1. The Limitations of Traditional Communication

Traditional communication systems, such as radio waves and fiber optics, rely on light-speed transmission, meaning that the speed of communication is limited by the distance that light can travel in a given time. This limitation becomes a major obstacle when considering interstellar communication or communication across vast distances in space. Currently, no method allows for real-time communication across such distances, making deep space missions or communication between distant star systems nearly impossible.

Zhang's theory offers a potential solution by focusing on the manipulation of space rather than relying solely on electromagnetic waves. The key lies in the concept of **cylindrical spiral motion of space**, which could theoretically allow information to be transmitted faster than light by bypassing the constraints of traditional communication methods.

2. Manipulating Space for Instantaneous Communication

In Zhang's theory, space is not a static vacuum but a dynamic entity in constant motion. The motion of space, particularly its cylindrical spiral pattern, creates opportunities for information to move faster than traditional light-speed communication would allow. By **manipulating the structure of space,** Zhang suggests that information could be transmitted through a process called **space displacement**, effectively compressing the distance between two points.

This form of communication would operate by altering the fabric of space itself, allowing signals to travel along these manipulated paths at speeds far exceeding that of light. Since the distance between two points is reduced to zero during light-speed travel (as covered in earlier chapters), it follows that **communication across those points could occur instantaneously**. By harnessing space manipulation, we could bypass the limitations imposed by traditional physics and enable **faster-than-light communication**.

3. Potential Applications and Implications

The ability to communicate across vast distances instantaneously would have profound implications, particularly in fields like space exploration and scientific research. Interstellar missions could maintain real-time communication with Earth, greatly enhancing the feasibility of long-distance space travel. Explorers on distant planets or in other star systems could relay their findings back to Earth without delay, allowing for **instantaneous data transmission**.

Moreover, faster-than-light communication would revolutionize the way we exchange information on Earth. Communication networks could be established that transmit information across the globe or even across solar systems without any delay, eliminating the bottlenecks imposed by current light-speed technologies.

4. Overcoming Energy Loss

One of the key challenges in long-distance communication is **energy loss** — the degradation of signal strength as it travels across vast distances. In Zhang's theory, manipulating space could also address this issue. By creating a controlled pathway through space for the information to travel, **minimal energy loss** would occur, resulting in much more efficient transmission of data.

In traditional communication systems, the further the signal has to travel, the more energy is lost, and the weaker the signal becomes. With space manipulation, the effective distance between two points is reduced, meaning that the energy required to transmit signals is greatly minimized. This would lead to more reliable communication systems with less energy expenditure, making faster-than-light communication not only faster but also more energy-efficient.

5. Conclusion

Zhang XiangQian's *Unified Field Theory* introduces the possibility of **faster-than-light communication** by utilizing the manipulation of space. By altering the cylindrical spiral motion of space and reducing the effective distance between two points, information could be transmitted almost instantaneously across vast distances. This breakthrough would revolutionize communication on Earth and enable real-time communication across star systems, marking a new era in human exploration and technology. The potential to overcome energy loss during transmission further highlights the practicality of this visionary approach to communication.

Chapter 8: Teleportation and Instant Travel

The concept of teleportation — the instantaneous transfer of objects or people from one location to another — has long been a staple of science fiction. From popular culture to futuristic visions of technology, teleportation represents the ultimate convenience in transportation, bypassing the limitations of speed, distance, and physical barriers. While current scientific understanding limits our ability to achieve this, Zhang XiangQian's *Unified Field Theory* offers a groundbreaking approach to teleportation through the manipulation of space itself.

1. The Limitations of Traditional Transportation

In the modern world, transportation is bound by physical laws, requiring energy to move objects from one location to another. Whether by car, plane, or rocket, all traditional forms of transportation are subject to the limitations of inertia, friction, and fuel consumption. Even at the speed of light, it would take years to travel between stars, making deep space exploration impractical using conventional methods.

Teleportation, however, eliminates the need for motion through space. By manipulating the structure of space, it becomes possible to instantaneously transport objects or people across vast distances without the need for intermediate travel. Zhang's theory provides a framework for understanding how **instant travel** could be achieved by altering the **cylindrical spiral motion of space** that surrounds all objects.

2. Space Manipulation and Instantaneous Travel

Zhang's theory is built on the idea that space is not a passive entity but is constantly in motion, radiating outward in a cylindrical spiral from every object at the speed of light. This motion of space can be manipulated to create **space superposition**, where the space surrounding two locations overlaps. When this happens, the effective distance between those two locations collapses to zero, allowing for **instantaneous travel**.

In the context of teleportation, this means that by manipulating the space between two points, we can reduce the distance to nothing, effectively moving an object from one location to another without the need for movement through physical space. This process would involve

reducing the mass of the object to zero, allowing it to bypass the constraints of inertia and motion. Once the object reaches the desired location, its mass would be restored, ensuring a safe and stable re-entry into normal space.

This two-stage process of mass reduction and mass restoration forms the foundation of Zhang's teleportation model, offering a practical mechanism for achieving **instant travel**.

3. Challenges of Teleportation

While the theory of space manipulation and teleportation is revolutionary, there are several challenges that need to be addressed. The first is the **precision of space manipulation**. In order to successfully teleport an object or person, the manipulation of space must be extremely precise, ensuring that the object arrives at the intended location without distortion or error.

Another challenge is **energy control**. While Zhang's theory suggests that the energy required for teleportation would be minimal once the process of mass reduction begins, the initial setup for manipulating space and achieving massless travel may require advanced technology that is not yet available. The challenge lies in finding a way to control the space around objects without causing unintended effects on the surrounding environment.

Finally, there is the issue of **mass restoration**. Once the object has been teleported to its destination, its mass must be restored in a controlled manner to avoid destabilization. This process requires careful handling to ensure that the object remains intact and unharmed during the teleportation process.

4. Potential Applications of Instant Travel

If the challenges of teleportation can be overcome, the implications for transportation are profound. Instant travel could eliminate the need for traditional vehicles and infrastructure. There would be no need for roads, planes, or even spacecraft, as people and objects could be transported instantaneously from one location to another. The energy and time savings alone would revolutionize how we think about transportation.

Space exploration would also be transformed. Instead of spending years traveling to distant planets or star systems, astronauts could be teleported instantly to their destination and back, allowing for near-instant exploration of the universe. This would open up new opportunities for scientific discovery and human expansion beyond Earth.

On Earth, teleportation could be used to transport goods, people, and resources across continents or even between planets without the need for traditional transportation methods. This would not only make travel faster but also reduce the environmental impact of transportation, as no fuel or physical infrastructure would be required.

5. The Future of Teleportation

While teleportation remains a speculative technology at present, Zhang's theory offers a pathway toward making it a reality. By manipulating space and reducing mass to zero, the limitations of traditional transportation can be bypassed, making **instant travel** a possibility. As technology advances and our understanding of space manipulation grows, the dream of teleportation could one day become a practical and efficient mode of transportation.

6. Conclusion

Teleportation, long considered a fantasy, may become a reality through Zhang XiangQian's *Unified Field Theory*. By harnessing the motion of space and reducing mass to zero, **instant travel** could revolutionize how we move across the planet, the solar system, and the universe. While challenges remain in terms of precision, energy control, and mass restoration, the potential benefits of teleportation are undeniable. This breakthrough would redefine transportation and open up a new era of possibilities for humanity.

Chapter 9: Artificial Fields and Virtual Structures

One of the most intriguing possibilities introduced by Zhang XiangQian's *Unified Field Theory* is the creation of **artificial fields** and **virtual structures** through the manipulation of space. In classical physics, fields such as gravitational and electromagnetic fields exist naturally, influencing the behavior of matter and energy. However, Zhang's theory suggests that by controlling the motion of space, we could create artificial fields with properties tailored to our needs, potentially allowing us to shape matter and energy in unprecedented ways.

1. What Are Artificial Fields?

In traditional physics, fields like gravity and electromagnetism are fundamental forces that influence how objects interact with each other. Gravitational fields, for instance, dictate how objects with mass attract one another, while electromagnetic fields govern the interactions between charged particles. These fields are considered natural phenomena that arise from the fundamental properties of space and matter.

Zhang's *Unified Field Theory*, however, presents the possibility of creating **artificial fields** — fields that are not a natural result of physical laws but are instead generated through the **manipulation of space**. By altering the **cylindrical spiral motion of space** surrounding objects, it is theoretically possible to create fields that mimic or even surpass the properties of natural fields.

For example, an **artificial gravitational field** could be generated around an object, enabling precise control over how that object interacts with its surroundings. Similarly, **electromagnetic fields** could be artificially created and manipulated to control energy flow or protect sensitive systems from radiation.

2. Creating Virtual Structures in Space

Beyond artificial fields, Zhang's theory introduces the concept of **virtual structures** — objects or systems created purely through the manipulation of space. These structures would not be composed of physical matter but would exist as **fields** that exhibit the properties of solid objects. In essence, virtual structures could serve as **energy-based frameworks** that interact with real-world matter in ways that traditional materials cannot.

Imagine a **virtual wall** made of space manipulation, capable of blocking or redirecting energy without the need for physical material. Such a structure could be used in applications ranging from advanced energy shields for spacecraft to containment fields for hazardous materials. These virtual structures could also serve as building blocks for advanced technologies, enabling us to create **dynamic, reconfigurable environments** that adapt to different needs.

For instance, in space exploration, virtual structures could be deployed to create temporary habitats or energy shields that protect astronauts from radiation. Unlike traditional materials, these virtual structures would be infinitely customizable, taking any shape or form depending on the requirements of the mission.

3. Applications of Artificial Fields

The creation of artificial fields would have significant implications across a range of scientific and technological fields. One of the most promising applications is in **energy control**. By generating artificial fields, we could develop more efficient systems for storing and transferring energy. **Artificial electromagnetic fields** could be used to create systems that transfer energy across long distances with minimal loss, leading to breakthroughs in **wireless energy transmission**.

In addition, **artificial gravitational fields** could revolutionize space exploration. Spacecraft could generate their own gravitational fields, allowing for stable movement through space without relying on traditional propulsion systems. These fields could also be used to create **artificial gravity** for space stations, providing astronauts with an Earth-like environment even in deep space.

On Earth, artificial fields could be employed in industries ranging from **construction** to **medicine**. Virtual structures could be used to create temporary, energy-based frameworks for building or repairing large structures without the need for traditional scaffolding. In the medical field, artificial fields could be used to create **non-invasive surgical tools** that manipulate tissues and organs without the need for physical contact.

4. The Implications for Science and Technology

The potential of **artificial fields** and **virtual structures** goes beyond practical applications. These breakthroughs challenge our understanding of the fundamental nature of matter and energy. If we can manipulate space to create fields and structures that function like physical objects, the line between **matter** and **energy** begins to blur. This raises profound questions about the nature of reality itself and opens the door to new avenues of scientific exploration.

Moreover, the ability to create **fields that interact with matter** without requiring physical materials could lead to advances in **quantum mechanics** and **particle physics**. By studying the behavior of matter within these artificial fields, scientists could gain new insights into the fundamental forces of the universe and uncover previously unknown phenomena.

5. Virtual Structures in Future Technologies

As technology advances, **virtual structures** could play a central role in the design of next-generation technologies. These energy-based structures could be used to create **virtual environments** for training, research, and entertainment. Instead of relying on physical components, engineers could design **virtual systems** that interact with the real world, offering more flexibility and efficiency.

For example, in **virtual reality** systems, artificial fields could be used to create physical sensations without the need for physical objects. Users could experience tactile feedback or interact with virtual objects in a way that feels real, even though the structures they are interacting with are composed entirely of manipulated space. This could revolutionize fields like **entertainment**, **education**, and **medical training**, where realistic simulations are essential.

6. Conclusion

Zhang XiangQian's *Unified Field Theory* opens up the possibility of creating **artificial fields** and **virtual structures** through the manipulation of space. These breakthroughs could revolutionize fields ranging from energy control to space exploration, allowing us to create **energy-based frameworks** that function like physical matter. As we explore the potential of these technologies, the line between **matter** and **energy** begins to blur, challenging our understanding of the universe and opening up a new frontier of scientific discovery.

Chapter 10: Body Replication and Consciousness Transfer

The concept of **body replication** and **consciousness transfer** has long intrigued scientists and philosophers alike, but Zhang XiangQian's *Unified Field Theory* and his experiences on **Planet Guoke** offer a revolutionary perspective on these topics. On Planet Guoke, **consciousness transfer** is not a far-off dream but a practical reality. The key to immortality on Guoke lies in separating a person's consciousness from their body, allowing it to be preserved and transferred to a new vessel when needed (Grace, 2025).

1. Body Replication on Planet Guoke

On Planet Guoke, **factories create bodies** without consciousness. These bodies are physically identical to humans, but they lack the defining essence of a person— their consciousness. This approach to body replication moves beyond the idea of cloning or genetic duplication, focusing instead on the **creation of new physical vessels** that can be occupied by a consciousness.

These bodies are essentially **empty shells**, waiting for a consciousness to inhabit them. When a person's original body deteriorates beyond repair, whether due to aging, illness, or injury, they can simply transfer their consciousness into a new body. This process eliminates the need for traditional medical interventions and ensures that individuals can **continue living indefinitely**, as long as their consciousness remains intact.

This form of **body replication** highlights a key principle from Zhang's experiences: **the physical body is not what defines a person**. In Guoke's society, the body is seen as a temporary vessel, interchangeable and replaceable. What makes a person unique—their thoughts, memories, and experiences—is their **consciousness** (Grace, 2025).

2. Consciousness Transfer: The Key to Immortality

The most profound aspect of Zhang's insights from Planet Guoke is the ability to **transfer consciousness** from one body to another, or even into a **computer system**. The people of Guoke believe that the majority of human traits are the same across individuals. What truly makes a person unique is their **consciousness**, the essence that carries their thoughts, memories, personality, and experiences. By preserving and transferring this consciousness, the people of Guoke have effectively achieved **immortality**.

In this model, a person's consciousness is transferred into a computer system, where it can be stored indefinitely. When the time comes to replace a person's body, the **consciousness is re-uploaded** into a new, fully functional body. This transfer process is seamless, ensuring that the individual retains all of their memories, personality traits, and sense of self.

The key to this process lies in Zhang's **Unified Field Theory**, which describes how consciousness is not confined to the brain or body but is instead a **spatial and energetic pattern**. By mapping this pattern and transferring it into a new vessel, the people of Guoke have overcome the limitations of mortality (Grace, 2025).

3. Challenges of Consciousness Transfer

While the technology on Planet Guoke is advanced enough to allow for **body replication** and **consciousness transfer**, these processes raise significant challenges when considered from a human perspective.

One major challenge is the concept of **continuity**. If a person's consciousness is transferred from one body to another, is it truly the same person, or is the original self lost in the process? On Guoke, this is not a concern, as the people believe that the essence of identity resides purely in the consciousness. However, from a human perspective, this raises questions about the **nature of self** and whether identity remains intact during the transfer.

Another challenge is the **technology required** to map and transfer consciousness. While the people of Guoke have developed the ability to store consciousness in computers and transfer it between bodies, the complexity of human consciousness — particularly the connections between neurons, memories, and experiences — remains an obstacle for current human science. Accurately capturing the full scope of a person's consciousness and transferring it without losing important elements of identity is a task that has yet to be fully understood or realized.

4. Ethical Implications

The idea of **consciousness transfer** also raises significant **ethical questions**. On Planet Guoke, where immortality has been achieved through this process, the society has adapted to the concept that the body is just a vessel. However, for humans, the notion of replacing bodies and preserving consciousness could lead to questions about **personal autonomy**, **individual rights**, and **what it means to be human**.

For instance, if consciousness can be preserved indefinitely and transferred into new bodies, would this lead to a new form of **class inequality**, where only those with access to the technology can achieve immortality? Additionally, if consciousness is stored in a computer, what rights would that consciousness have? Would it be treated as a living person, or would it become a **digital entity** without the same rights and freedoms?

5. Future Implications of Consciousness Transfer

If the challenges of **consciousness transfer** can be overcome, the potential applications are staggering. One of the most obvious is the ability to **extend human life indefinitely**. By transferring consciousness into new bodies, individuals could avoid the limitations of aging and physical deterioration.

In addition to life extension, **consciousness transfer** could also play a key role in **space exploration**. Human consciousness could be transferred into **synthetic bodies** designed to withstand the harsh conditions of space travel. These synthetic bodies, built without the need for biological functions, could survive in extreme environments such as deep space, allowing humans to explore the universe without the need for life support systems.

Finally, **consciousness transfer** could reshape our understanding of **virtual reality**. By uploading consciousness into a computer, individuals could experience fully immersive digital worlds that feel as real as physical reality. This could revolutionize entertainment, education, and even social interaction, creating entirely new ways for people to connect and experience the world.

6. Conclusion

Zhang XiangQian's experiences on **Planet Guoke** reveal a society where **body replication** and **consciousness transfer** have become the key to immortality. By preserving a person's consciousness and transferring it to new bodies, the people of Guoke have overcome the limitations of mortality and redefined the relationship between the physical body and the self. While significant challenges remain in terms of technology and ethics, these insights offer a glimpse into a future where **consciousness is no longer confined to the body**, opening up new possibilities for life extension, space exploration, and the nature of human existence.

Part 4: Unifying the Forces of Nature

Chapter 11: Understanding the Four Forces

Zhang XiangQian's *Unified Field Theory* presents a revolutionary understanding of the four forces — **gravity, electric force, magnetic force**, and **nuclear force** — by unifying them through the **cylindrical spiral motion of space**. Zhang's theory challenges traditional models by providing mathematical proofs and experimental validation that show how these forces are different manifestations of the same underlying principle: the motion of space. Zhang's work also directly builds on, and at times challenges, several established theories in modern physics.

1. Limitations of Current Unified Theories

In contemporary physics, the four forces — **gravity, electromagnetic force, strong nuclear force**, and **weak nuclear force** — are treated as distinct. While progress has been made in unifying some forces, others, particularly gravity, remain disconnected. Key areas where Zhang's theory advances beyond existing models include:

1) Einstein's General Theory of Relativity (Gravity)

Einstein's theory describes gravity as the curvature of spacetime caused by mass. However, it treats gravity as distinct from the quantum forces (electromagnetic, strong nuclear, and weak nuclear). Zhang's theory goes beyond Einstein by showing that **gravity** can be explained by the **spiral motion of space**, unifying it with the other forces.

2) Electroweak Theory (Unification of Electromagnetic and Weak Forces)

While the **Standard Model** unifies the **electromagnetic** and **weak nuclear forces** under the **electroweak theory**, gravity and the strong nuclear force are left outside. Zhang's theory demonstrates that **gravity** and **electromagnetic forces** are directly connected through the motion of space, providing a clearer pathway for unification that current models don't fully address.

3) Quantum Chromodynamics (QCD) and the Strong Nuclear Force

QCD explains how the strong force holds quarks together inside protons and neutrons, but it doesn't unify this force with gravity or electromagnetism. Zhang's theory proposes that the **nuclear force** arises from space's behavior at the subatomic level, thus offering a unified perspective on the strong nuclear force.

4) String Theory

String theory is a speculative attempt to unify all four forces by suggesting that particles are one-dimensional "strings" rather than points. However, it lacks experimental validation and remains highly abstract. Zhang's theory provides a more grounded explanation, with concrete **experimental proofs** demonstrating the unification of forces through space manipulation, overcoming the speculative nature of string theory.

2. Zhang's Unified Theory: Overcoming Limitations

Zhang XiangQian's theory provides mathematical proofs and experimental evidence to show how **gravity**, **electric force**, **magnetic force**, and **nuclear force** are interconnected. Unlike traditional models, which leave gravity as an outlier, Zhang's theory unifies these forces by showing that they all arise from the **cylindrical spiral motion of space**.

1) Unifying Gravity and Electromagnetism

Zhang's theory shows that the interaction between charged particles (electric force) and magnetic fields (magnetic force) is linked to gravity through space's motion. This explanation goes beyond Einstein's **General Relativity** by providing a mechanism that unifies gravity with the quantum forces.

2) Mathematical Proofs

Zhang has developed mathematical proofs that align with established theories but go further in demonstrating how space's motion governs all four forces. These proofs show that gravity, electric force, magnetic force, and nuclear force can all be understood as **different manifestations of space's behavior**.

3) Experimental Validation

Zhang's patented experiments provide tangible proof of his theory's claims:

i. **Linear Gravitational Field from Accelerating Positive Charges**: This experiment showed that accelerating positive charges generates a linear gravitational field, directly linking **electromagnetism and gravity**—a connection that **string theory** and

General Relativity haven't experimentally demonstrated.

ii.   **Vortex Gravitational Field from Changing Magnetic Fields**: This experiment demonstrated that changing magnetic fields can generate a vortex-like gravitational field, further proving the connection between **magnetic forces** and **gravitational fields**.

These experiments, which have been patented, offer concrete evidence for the unification of forces, distinguishing Zhang's theory from speculative models like **string theory** (Zhang, 2025).

3. Specific Theories Proven by Zhang's Unified Field Theory

Zhang's *Unified Field Theory* aligns with and extends beyond the following key established theories:

1) Electromagnetic Theory

Zhang's theory expands on **Maxwell's equations** by showing how the manipulation of space affects electromagnetic fields, demonstrating that **electric** and **magnetic forces** are interconnected with gravity (Maxwell, 1865).

2) General Relativity

While Zhang's theory incorporates aspects of **Einstein's curvature of spacetime**, it goes further by showing that gravity itself is the result of space's cylindrical spiral motion, providing a direct connection to the other forces (Einstein, 1915).

3) Quantum Mechanics

Zhang's theory extends **quantum mechanics** by explaining how space's motion influences forces at both quantum and cosmic scales, offering a unification of quantum forces with gravity (Planck, 1900).

Zhang's mathematical equations, backed by **patents** and **experimental validation**, provide a framework that aligns with these established theories while offering a more cohesive and unified explanation of the four forces (Zhang, 2025).

4. Zhang's Four Forces

Zhang's theory identifies four key forces:

1) Gravity

Governed by the spiral motion of space around massive objects.

2) Electric Force

Generated by the interaction of charged particles and space.

3) Magnetic Force

Linked to moving charges and the space they influence.

4) Nuclear Force

The force binding atomic nuclei, which Zhang shows to be influenced by space's behavior at a subatomic level.

Each of these forces is a manifestation of the **cylindrical spiral motion of space**, which Zhang argues is the true unifying principle behind all forces in the universe (Zhang, 2025).

5. Conclusion

Zhang XiangQian's *Unified Field Theory* provides a unified explanation of **gravity, electric force, magnetic** force, and **nuclear force** through the **cylindrical spiral motion of space**. His mathematical proofs align with and extend current theories, including **General Relativity**, **Maxwell's electromagnetic theory**, and **quantum mechanics**. Zhang's theory is supported by **patented experiments**, offering tangible evidence that overcomes the limitations of speculative theories like **string theory** and providing a more cohesive and experimentally supported understanding of the forces governing the universe.

Chapter 12: How Gravity and Space Are Linked

In Zhang XiangQian's *Unified Field Theory*, **gravity** is not a mysterious force acting at a distance, as traditionally understood, but is directly linked to the **cylindrical spiral motion of space**. This relationship redefines how we understand gravitational fields and their interaction with matter and energy, offering new possibilities for controlling gravity through the manipulation of space.

1. Gravity in Traditional Physics

In **Einstein's General Theory of Relativity**, gravity is explained as the curvature of spacetime caused by the presence of mass (Einstein, 1915). Massive objects, like planets and stars, warp the spacetime around them, creating the force of gravity that pulls objects toward them. While this explanation works well on a macroscopic scale, it does not unify gravity with the other fundamental forces, particularly at the quantum level.

Moreover, the traditional view of gravity lacks a direct mechanism for how gravity interacts with the quantum world. This leaves a significant gap between **General Relativity** and **quantum mechanics**, which has led to numerous efforts to develop a **quantum theory of gravity**— none of which have been conclusively proven.

2. Zhang's Reinterpretation: The Role of Space in Gravity

Zhang's *Unified Field Theory* bridges this gap by explaining that **gravity is the result of the spiraling motion of space** around massive objects. In Zhang's model, space is not static but is constantly in motion, radiating outward in a cylindrical spiral. This motion of space creates what we observe as gravitational fields, and the strength of the gravitational field is directly related to the intensity of space's spiral motion.

According to Zhang, **gravity** is not just the result of mass bending spacetime, but a consequence of how space itself behaves in the presence of mass. When a massive object, such as a planet or star, is present, the surrounding space is drawn into a **spiral pattern**, pulling objects toward the center of the mass. The motion of space replaces the need for "action at a distance," providing a **local, dynamic mechanism** for how gravitational fields operate (Zhang, 2025).

3. Zhang's Experiments on Gravity

Zhang's theory of gravity is supported by experimental evidence from his patented work:

1) Linear Gravitational Field from Accelerating Positive Charges

This experiment demonstrated that accelerating positive charges in a high-voltage environment generates a **linear gravitational field**. This finding shows that gravitational effects can be produced by manipulating the motion of space around charged particles, connecting **electromagnetism** with gravity.

2) Vortex Gravitational Field from Changing Magnetic Fields

In this experiment, Zhang generated a **vortex gravitational field** by changing magnetic fields in a vacuum chamber. This demonstrated that gravitational fields can be influenced by **magnetic forces**, further proving the link between space's motion, electromagnetism, and gravity.

These experiments show that **gravity** can be controlled by manipulating the motion of space, offering experimental proof for Zhang's unified view of the forces.

4. Implications of Zhang's Gravity-Space Link

The idea that **space itself moves** and generates gravitational fields opens up numerous possibilities for both theoretical physics and practical applications:

1) Artificial Gravity

By manipulating space's motion, it may be possible to create **artificial gravitational fields**. This could revolutionize space exploration, where spacecraft or space stations could generate their own gravity, allowing astronauts to experience Earth-like conditions while in orbit or traveling to distant planets.

2) Gravity Control for Propulsion

Zhang's theory suggests that controlling the motion of space could be used for advanced propulsion systems. By generating specific gravitational fields, it might be possible to propel objects through space without the need for traditional fuels, drastically improving the efficiency of space travel.

3) Energy Applications

The ability to manipulate space and gravity could lead to new forms of energy generation. **Gravitational fields** could be harnessed to produce energy, or artificial fields could be used to move objects and materials in ways that current technologies cannot.

5. Gravity and the Quantum World

Zhang's theory also offers new insights into how gravity interacts with the quantum world. Traditional theories have struggled to reconcile gravity's behavior on large scales with the forces operating at the quantum level. However, Zhang's model provides a unified explanation by showing that **gravity** and **quantum forces** are both governed by the motion of space.

In this view, quantum particles interact with space's cylindrical spiral motion in the same way that larger objects do, but on a much smaller scale. This explains how gravity can affect quantum particles without the need for complex theoretical constructs like those in **quantum gravity** models.

6. Conclusion

Zhang XiangQian's *Unified Field Theory* offers a groundbreaking reinterpretation of **gravity**, showing that it is intrinsically linked to the **motion of space**. His theory, supported by experimental evidence, provides a unified framework for understanding how gravity interacts with other forces, while also opening the door to exciting technological applications, from artificial gravity to advanced propulsion systems. By rethinking gravity as a consequence of space's motion, Zhang's theory addresses key gaps in traditional physics and presents a new direction for understanding the forces that govern our universe.

Chapter 13: Energy, Forces, and the Future of Physics

In Zhang XiangQian's *Unified Field Theory*, the relationship between energy and forces is unified through a single equation that accounts for momentum changes due to added mass or changes in velocity. This unified equation provides a comprehensive framework that links the **electric**, **magnetic**, **gravitational**, and **nuclear forces** to the behavior of space and matter.

1. Unified Equation for Forces

The key equation in Zhang's *Unified Field Theory* is:

$$\vec{F} = \frac{d\vec{p}}{dt} = \vec{c}\,\frac{dm}{dt} - \vec{v}\,\frac{dm}{dt} + m\,\frac{d\vec{c}}{dt} - m\,\frac{d\vec{v}}{dt}$$

This equation expresses the total force $\vec{F}$ as the rate of change of momentum $p^{\rightarrow}$ over time t, incorporating both **mass changes** and **velocity changes**. It ties together multiple forces into a unified framework, showing how each force arises from different aspects of mass and velocity interactions with space (Zhang, 2025).

2. Interpretation of Forces in Unified Field Theory

The terms in this equation represent different forces within the unified framework:

1. $(\vec{c} - \vec{v}) \frac{dm}{dt}$: This term represents the force due to added mass, specifically **mass-change motion**. It captures how changes in an object's mass generate force.

2. $m \frac{d\vec{c} - d\vec{v}}{dt}$: This term represents the **acceleration force**, or the force due to changes in velocity.

These terms encompass the four fundamental forces in Zhang's *Unified Field Theory*:

1. **Electric Field Force**: $\vec{c} \frac{dm}{dt}$, where changes in mass interact with the speed of light $\vec{c}$, creating an **electric field**.

2. **Magnetic Field Force**: $\vec{v} \frac{dm}{dt}$, where mass changes interact with velocity $\vec{v}$, generating a **magnetic field**.

3. **Gravitational (Inertial) Force**: $m \frac{d\vec{v}}{dt}$, where changes in velocity create a **gravitational (inertial) force**.

4. **Nuclear Force**: $m \frac{d\vec{c}}{dt}$, where changes in the speed of light $\vec{c}$ result in the **nuclear force**.

This unified equation shows how different forces emerge from the fundamental interaction between mass, velocity, and space. It provides a new understanding of how energy and forces interact, demonstrating that they are all part of the same dynamic system governed by space's motion.

3. Energy and Forces in Zhang's Theory

Zhang's theory links energy and forces through the motion of space. The total energy in a system is the result of how space interacts with objects, and the forces acting on those objects arise from space's motion. Unlike traditional physics, which treats energy and forces separately, Zhang's unified equation shows that they are fundamentally connected.

For instance, the **electric force** arises from changes in mass as it interacts with space at the speed of light, while the **magnetic force** is generated by the interaction between mass changes and velocity. **Gravitational forces** are linked to changes in velocity, while **nuclear forces** are connected to the changes in the speed of light in space.

4. Technological Applications

This unified understanding of energy and forces opens up new possibilities for technological advancements:

1) Energy Generation

By manipulating the motion of space, new methods of **energy generation** could be developed, harnessing the interactions between mass, velocity, and forces to produce energy more efficiently.

2) Propulsion Systems

The ability to control forces like gravity and electromagnetism through space's motion could lead to new **propulsion technologies** that do not rely on traditional fuels, enabling faster and more energy-efficient space travel.

3) Wireless Power Transmission

The manipulation of space and forces could lead to advanced **wireless power transmission systems**, allowing for energy to be transmitted over long distances without the need for wires or infrastructure.

4) Non-invasive Medical Technologies

By controlling forces at a molecular level, it could be possible to develop **non-invasive medical technologies** that heal or repair tissues using field-based methods, without the need for physical surgery.

5. Conclusion

Zhang XiangQian's *Unified Field Theory* ties together the four fundamental forces—**electric, magnetic, gravitational**, and **nuclear**—through the unified equation $\vec{F} = \frac{d\vec{p}}{dt}$. This equation demonstrates how forces and energy are linked to the motion of space, providing a cohesive framework that opens the door to revolutionary technological advancements. From energy generation to propulsion and medical technologies, Zhang's unified understanding of forces and energy offers a new way of thinking about the future of physics.

Part 5: The Future of Technology and Science

Chapter 14: Space Travel, Energy, and the Future

Zhang XiangQian's *Unified Field Theory* challenges traditional approaches to space travel and energy production by proposing that we can harness the **cylindrical spiral motion of space** to propel spacecraft and generate energy. These advancements have the potential to revolutionize both the exploration of space and how we power our world.

1. Transforming Space Travel

Traditional propulsion systems, dependent on fuel and thrust, face significant limitations. Rockets must carry large amounts of fuel, and the energy required to overcome gravity and inertia is immense. Zhang's theory offers a new pathway—**manipulating the motion of space itself** to enable travel at unprecedented speeds.

1) Light-Speed Travel through Mass Reduction

At the core of Zhang's theory is the idea that by reducing an object's **mass to zero**, we can achieve **instantaneous light-speed travel**. This concept is based on the relationship between mass and space: when mass approaches zero, the distance between two points effectively collapses to zero. In practical terms, this would allow spacecraft to traverse vast distances in an instant, bypassing the need for traditional fuel or slow acceleration.

2) Propulsion via Space Manipulation

Beyond light-speed travel, Zhang's theory suggests that by controlling the **cylindrical spiral motion of space**, spacecraft could be propelled without the need for traditional fuel. This method would involve creating localized gravitational fields or harnessing space's natural motion to propel vehicles, offering a far more efficient means of travel across the universe.

3) Artificial Gravity

Long-term space travel poses challenges, including the lack of gravity, which can lead to serious health issues for astronauts. Zhang's theory suggests that by manipulating space, we could generate **artificial gravitational fields** on spacecraft, allowing travelers to experience Earth-like conditions even in deep space.

2. New Energy Solutions

The potential applications of Zhang's theory extend far beyond space travel. His ideas could also transform how we produce and distribute energy, offering new methods that bypass the limitations of current technologies.

1) Gravitational Energy Systems

By manipulating gravitational fields, it may be possible to develop new forms of **gravitational energy**. This energy could be harnessed through artificial gravitational wells or by utilizing space's spiral motion to create a sustainable and powerful energy source that doesn't rely on fossil fuels.

2) Electromagnetic Energy

Zhang's theory also suggests new ways to generate **electromagnetic energy**. By controlling the interactions between space and charged particles, we could create more efficient systems for producing and transmitting energy, possibly providing a more sustainable alternative to current electrical grids.

3) Space-Based Energy Solutions

With advancements in space travel, Zhang's theory opens the door to **space-based energy generation**, such as placing solar or gravitational energy systems in orbit or on distant planets. These systems could then transmit energy back to Earth or other planets, creating a decentralized, sustainable energy network.

3. Overcoming the Limits of Current Physics

One of the most significant contributions of Zhang's theory is its ability to overcome the limitations imposed by traditional physics and energy systems. Current models for space travel and energy production are constrained by the need for massive amounts of fuel and finite resources. Zhang's approach, however, shows that by manipulating space itself, we can unlock **new forms of propulsion** and **energy generation** that bypass these limitations entirely.

1) Sustainable Energy

Zhang's theory offers a pathway to developing energy systems that are both sustainable and efficient. By tapping into the natural forces of space, we can generate energy without depleting resources or harming the environment.

2) Advanced Space Technologies

As we enter a new era of space exploration, Zhang's theory provides the foundation for **advanced space technologies** capable of overcoming the challenges of gravity, propulsion, and long-distance travel. This opens up possibilities for colonizing other planets or traveling to distant star systems.

4. The Future of Human Civilization

The technological advancements enabled by Zhang's *Unified Field Theory* could have profound implications for the future of human civilization. As we gain mastery over the forces of space, we can solve pressing global challenges and push the boundaries of what is possible for humanity.

1) Interstellar Colonization

Zhang's theory provides the technological foundation for **interstellar colonization**, where humans could establish permanent settlements on other planets or moons. With light-speed travel and artificial gravity, long-term habitation of distant worlds becomes feasible.

2) A New Energy Paradigm

The development of sustainable energy systems based on Zhang's principles could usher in a **new global energy paradigm**, where energy is abundant, clean, and available to everyone. This would drastically reduce the world's reliance on fossil fuels and centralized energy grids.

3) Technological Prosperity

Mastering the forces of space could lead to a new era of **technological prosperity**, where innovations in space travel, energy generation, and other fields transform the way we live, work, and explore the universe.

5. Conclusion

Zhang XiangQian's Unified Field Theory presents a bold vision for the future of space travel and energy production. By unifying the forces of nature and showing how they emerge from the motion of space, Zhang's theory provides the foundation for revolutionary advancements in technology. From instantaneous light-speed travel to sustainable energy systems, Zhang's ideas offer a pathway to overcoming the limitations of current physics and unlocking new possibilities for the future of human civilization.

Chapter 15: The Future of Science and Technology

Building on the foundation of Zhang XiangQian's *Unified Field Theory*, the next frontier is the full-scale implementation of **Artificial Field Scanning Technology**. This chapter explores how this technology could revolutionize not only energy systems but also offer new ways to store information, manipulate consciousness, and redefine the limits of human exploration.

1. Artificial Fields and Energy Systems

Zhang's theory suggests that artificial gravitational fields, created through the manipulation of **electromagnetic fields**, could replace traditional forms of energy. By converting field energy, we could develop a **decentralized energy network** that no longer depends on fossil fuels or current electrical grids.

1) Solar Energy Concentration

Artificial fields could compress space, concentrating solar energy from vast areas, solving the global energy crisis.

2) Wireless Power Systems

Energy could be transmitted over great distances without the need for a physical grid, providing power to remote or underdeveloped areas.

2. Information Storage and Virtual Worlds

Artificial fields have the ability to compress space infinitely, which opens the door to limitless information storage. This breakthrough would revolutionize **data technologies**, allowing for the storage and transmission of massive datasets, far surpassing today's limits.

1) Infinite Information Storage

The compression of space could store vast amounts of information, enabling advanced data technologies.

2) Virtual Construction

Artificial fields could create **virtual objects** and environments, allowing multiple users to interact with a shared digital space.

3. Manipulating Consciousness and Immortality

One of the most extraordinary aspects of Zhang's theory is the potential to manipulate human **consciousness**. According to Zhang, consciousness is linked to the motion of space, and by controlling this motion, we could transfer or store human consciousness independently of the body.

1) Consciousness Storage

Artificial fields could read and store human consciousness, offering a path to preserving memories or even achieving **immortality**.

2) Consciousness Transfer

In the future, people could transfer their consciousness into different bodies or systems, allowing for a form of life extension that is currently science fiction.

4. Steps for Development and Implementation

The full potential of Artificial Field Scanning Technology requires further research and experimentation. Zhang outlines several key steps for its realization:

1) Refining Field Interaction Equations

Experiments have already demonstrated that accelerating charges can generate gravitational fields. The next step is refining the equations that govern the interaction between **electromagnetic fields** and **gravitational fields**.

2) Software and System Development

Specialized software must be developed to control artificial fields for applications in **energy systems**, **space exploration**, and **medical technology**.

3) Expanding Use Across Industries

Once the system is refined, artificial field scanning could replace current electrical grids and serve as the foundation for **space-based exploration** and **information technologies**.

5. Conclusion

Zhang XiangQian's *Unified Field Theory* offers a glimpse into a future where the fundamental forces of nature are fully understood and controlled. Through **Artificial Field Scanning Technology**, the boundaries of science are pushed beyond what was previously imagined, leading to a future where energy, consciousness, and space travel are no longer constrained by traditional limits. The development of this technology has the potential to redefine humanity's place in the universe, offering endless possibilities for exploration, energy, and even the nature of life itself.

Conclusion

Zhang XiangQian's *Unified Field Theory* offers a revolutionary framework for understanding the universe. By unifying the fundamental forces through the motion of space, Zhang provides a new lens for exploring not only the nature of gravity, electromagnetism, and nuclear forces but also the potential for breakthroughs in energy, transportation, and consciousness.

The practical applications of this theory, such as **Artificial Field Scanning Technology**, demonstrate that the future of science and technology could be shaped by a deeper understanding of space itself. Zhang's work challenges the limits of traditional physics, offering a vision for a future where the manipulation of space leads to innovations that were once thought impossible, from instantaneous travel to infinite energy.

As we move forward, Zhang's insights encourage us to reimagine the boundaries of science, inviting collaboration and further exploration. The true potential of his *Unified Field Theory* is just beginning to be realized, and its implications for the future are vast.

Appendices

Appendix A: Key Equations of Unified Field Theory

1) Mass Definition Equation

$$m = k \frac{\oiint dn}{\oiint d\Omega} = k \frac{n}{4\pi}$$

- **Explanation**: This equation defines the mass mmm at a point as a result of the distribution of spatial displacement vectors (nnn) surrounding that point in a solid angle of 4π4\pi4π. This shows that mass is derived from the interaction between space and the object, rather than being an intrinsic property of the object itself.

2) Unified Force Equation

$$\vec{F} = \frac{d\vec{p}}{dt} = \vec{c}\frac{dm}{dt} - \vec{v}\frac{dm}{dt} + m\frac{d\vec{c}}{dt} - m\frac{d\vec{v}}{dt}$$

- **Explanation**: This equation describes how momentum changes due to mass or velocity changes. It unifies the four fundamental forces in Zhang's theory:

    o **Electric Force:** $\vec{c}\frac{dm}{dt}$

    o **Magnetic Force:** $\vec{v}\frac{dm}{dt}$

    o **Gravitational (Inertial) Force:** $m\frac{d\vec{v}}{dt}$

○ **Nuclear Force**: $m \frac{d\vec{c}}{dt}$

3) Zhang's Signature Equation

$$\overrightarrow{p_{motion}} = m\,(\vec{c} - \vec{v})$$

- **Explanation**: This is the key momentum equation in Zhang's theory, where the momentum of an object is defined by its mass and the difference between the speed of light c and its velocity v. The second part represents the rest momentum when velocity is zero.

Appendix B: Zhang's Key Experiments

1) Linear Gravitational Field from Accelerating Positive Charges

- **Setup**: Positive charges were accelerated in a high-voltage environment, producing a linear gravitational field.

- **Results**: The gravitational field caused lightweight objects to move toward the positive terminal, demonstrating the ability to manipulate space through charged particles.

- **Significance**: This experiment supports Zhang's theory that gravitational fields can be controlled through electromagnetic interactions.

2) Vortex Gravitational Field from Changing Magnetic Fields

- **Setup**: Two spiral coils were placed around a vacuum chamber, and changing magnetic fields generated a vortex gravitational field.

- **Results**: A polyethylene ball inside the chamber began to rotate without external forces, proving the field's ability to alter motion.

- **Significance**: This experiment shows that space's behavior can be controlled by manipulating magnetic fields, further validating Zhang's theory.

Appendix C: Comparison with Mainstream Physics

1) General Relativity vs. Unified Field Theory

- **General Relativity**: Gravity is the result of spacetime curvature caused by mass and energy.

- **Unified Field Theory**: Gravity is a product of space's cylindrical spiral motion around objects, which affects the motion of all forces.

- **Advancement**: Zhang's theory provides a more dynamic explanation, allowing gravity to be manipulated through electromagnetic fields.

2) Quantum Mechanics and Zhang's Theory

- **Quantum Mechanics**: Describes forces at the smallest scales without including gravity.

- **Unified Field Theory**: Unifies quantum forces and gravity by demonstrating that both arise from the behavior of space. This provides a more cohesive view, linking phenomena from the microscopic to the cosmic level.

Appendix D: Diagrams and Visual Aids

Diagram 1: Cylindrical Spiral Motion of Space

A visual representation of how space moves around objects in a spiral pattern, creating the forces we observe.

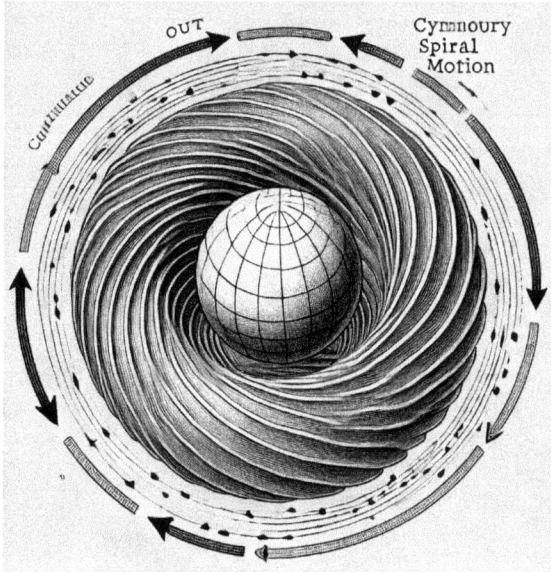

Diagram 2: Artificial Field Scanning Experiment Setup

A diagram showing the layout of Zhang's vortex gravitational field experiment, illustrating how magnetic fields are manipulated to generate artificial gravitational fields.

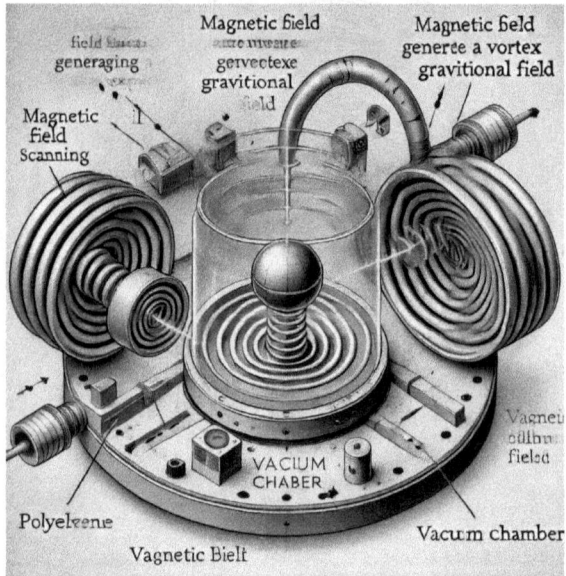

Diagram 3: Space Superposition

A visual aid explaining how the superposition of spaces from different objects leads to mass reduction and instantaneous light-speed travel.

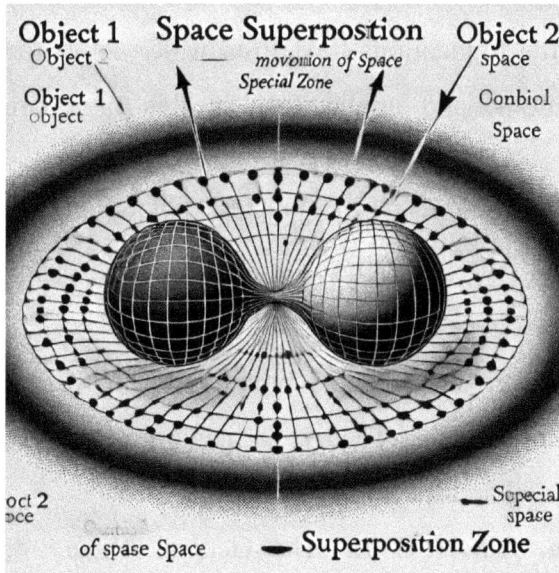

References

Zhang, XiangQian. 2025. Unified Field Theory (Academic Edition) – Extraterrestrial Technology. Hope Grace Publishing.

This is the foundational text from which all of the concepts, equations, and applications in this book are derived.

Grace, Hope. 2025. Voyage Throughout Guoke Planet: Inspired by the True Experiences of Zhang XiangQian. Hope Grace Publishing.

This book covers Zhang's broader vision for the application of Unified Field Theory, particularly in relation to interstellar travel and space technology.

Einstein, Albert. 1915. The General Theory of Relativity.

Cited for comparison between Einstein's classical understanding of gravity as the curvature of spacetime and Zhang's reinterpretation of gravity as the cylindrical spiral motion of space.

Maxwell, James Clerk. 1865. A Dynamical Theory of the Electromagnetic Field.

Cited for expanding upon electromagnetic theory and demonstrating how space affects electromagnetic fields.

Planck, Max. 1900. Quantum Theory of Energy Quantization.

Referenced to explain how Zhang's Unified Field Theory aims to unify quantum mechanics with gravity.

Glossary

Artificial Field Scanning: A technology that uses positive and negative gravitational fields generated by changing electromagnetic fields. This allows for the manipulation of space, time, and matter through computer-controlled systems, enabling various applications such as energy transmission, transportation, and material control.

Cylindrical Spiral Motion: A key concept in Zhang's *Unified Field Theory* describing how space moves outward from objects in a continuous, spiral pattern. This motion underlies the creation of forces like gravity, electromagnetism, and nuclear forces.

Electromagnetic Fields: Fields generated by electrically charged particles, which exert forces on other charged particles. In Zhang's theory, these fields interact with gravitational fields to form the basis of his *Artificial Field Scanning Technology*.

Gravitational Field: A field surrounding an object with mass, caused by the cylindrical spiral motion of space around the object. In Zhang's theory, gravitational fields can be manipulated by changing electromagnetic fields, offering control over gravitational forces.

Mass Definition Equation: $m = k \dfrac{\oiint dn}{\oiint d\Omega} = k \dfrac{n}{4\pi}$,

an equation that defines mass as the distribution of spatial displacement vectors around a point, showing mass as a product of space rather than an intrinsic property.

Momentum (Zhang's Signature Equation): $\overrightarrow{p_{motion}} = m\,(\vec{c} - \vec{v})$, representing the momentum of an object as determined by its mass and the difference between the speed of light and the object's velocity. The rest momentum is given by $\overrightarrow{p_{rest}} = m'\,\vec{c}$.

Quantum Mechanics: A branch of physics that deals with the behavior of particles at the smallest scales. Zhang's theory unifies quantum mechanics and gravity by showing both as consequences of space's motion.

Rest Momentum: The momentum of an object when its velocity is zero. In Zhang's theory, rest momentum is a key factor in determining how mass and energy interact with space.

Space Superposition: A concept where the spaces surrounding different objects overlap, effectively reducing the mass of the objects involved. This superposition can lead to instantaneous light-speed travel, as described in Zhang's theory.

Unified Force Equation: $\vec{F} = \frac{d\vec{p}}{dt} = \vec{c}\,\frac{dm}{dt} - \vec{v}\,\frac{dm}{dt} + m\,\frac{d\vec{c}}{dt} - m\,\frac{d\vec{v}}{dt}$, an equation that describes how the forces in the universe (electric, magnetic, gravitational, and nuclear) are all manifestations of the same underlying principles governed by space's motion.

Vortex Gravitational Field: A gravitational field generated by changing magnetic fields, demonstrated in Zhang's experiment using spiral coils. This phenomenon shows how electromagnetic fields can directly influence gravitational forces, providing the basis for *Artificial Field Scanning Technology*.

www.ingramcontent.com/pod-product-compliance
Lightning Source LLC
Chambersburg PA
CBHW071710210326
41597CB00017B/2420